The author expressed her appreciation to Brian H. Mason, Ph.D., who was curator of the Physical Geology and Minerology for the American Museum of Natural History, New York, for his suggestions and contributions to the text and illustrations contained in the book.

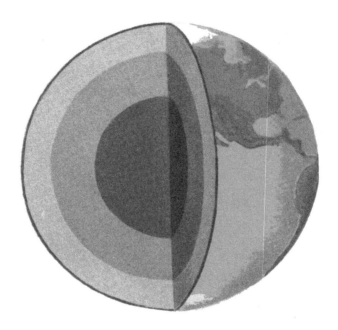

This book was originally published in 1958 by Franklin Watts, Inc.

This edition has been published by Living Library Press ©2018.

No part may be reproduced without the publisher's permission.

ISBN #: 978-0-692-09071-8

THE FIRST BOOK OF
THE EARTH

By O. IRENE SEVREY

Pictures by MILDRED WALTRIP

Living Library Press
P.O. Box 16141
Bristol, Virginia 24209

RELIEF MAP OF THE UNITED STATES — the darker areas show high land.

contents

1. our planet — 1
2. how did it begin? — 6
3. what is the earth made of? — 11
4. the rock families — 16
5. the earth's treasures — 22
6. bending crust — 23
7. volcanoes — 29
8. hot rock underground — 31
9. down come the mountains — 36
10. a river's story — 39
11. the earth's history book — 42
12. ground water — 47
13. glaciers — 52
14. the earth through the ages — 55
    index — 63

our planet

Did you ever look at a map made especially to show the rivers and mountains, the lakes and plains—the "up-hills and down-dales" of our land? It is called a relief map, and it is quite different from the usual map that shows countries and cities. On it you can really see the hills rising from the flat land. And you can see the rivers beginning as narrow streams and growing wider and wider as they run toward the sea. Some of the land is spotted all over with lakes, large and small. Other parts of the country, the deserts, have no waterways at all.

Suppose someone were to ask you to make a quick, very rough map something like this, showing the little hills and valleys, the brooks and ponds of your own small neighborhood. You could probably do it from memory.

"How do you know that's the way it is, without even looking?" a friend might ask.

"Why, that's the way it's always been. I've climbed that hill dozens of times," you might possibly answer, pointing to a hill you have made on your map.

Yet all those hills and valleys, brooks and ponds, and even the earth itself, had a beginning. Did you ever stop to wonder how they all happened to be exactly as they are? Why are there mountains on one part of our earth, plains on another, oceans in another? Why are there rocks, lakes, swamps, and caves?

Men puzzled over questions like these for thousands of years. More and more they looked closely at the earth itself. And there they gradually worked out the answers to many of their questions.

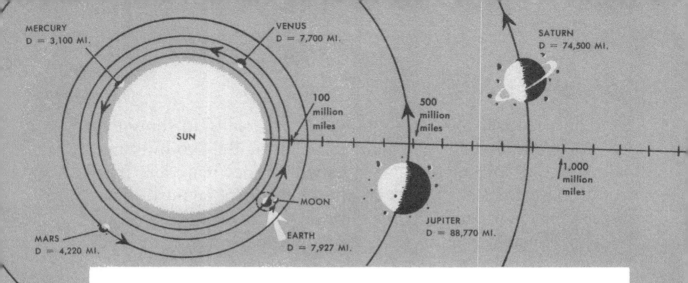

The earth is a picture book that tells much of its story to those who know how to read it.

The men who can read our earth best are the geologists. They study the earth. Another group of scientists, the astronomers, tell us about the stars and the earth's place among them. The geologists and the astronomers between them have pieced together much of the story of the earth.

They are agreed that the earth is a huge ball of rock moving at great speed around our sun. It is called a planet, a name that means "wandering." The sun is circled by nine planets—four that are larger than the earth, three that are smaller, and one, Venus, that is much the same size.

Some of the planets have moons—balls of rock that travel with them. The earth has only one moon, while Jupiter, another planet, has twelve.

Very small balls of rock called *asteroids*, journey along with the planets. Thousands of small, broken pieces of rock whirl along with them, too. These are called meteors from a Greek word meaning, "high in the sky." Some astronomers think meteors may

2

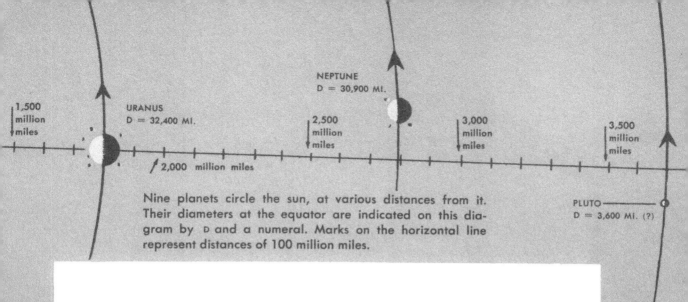

Nine planets circle the sun, at various distances from it. Their diameters at the equator are indicated on this diagram by D and a numeral. Marks on the horizontal line represent distances of 100 million miles.

have come from a planet that exploded. Meteors which fall into the earth's atmosphere usually become red hot and burn up before they reach the ground. Often, if you watch at night, you can see them streaking across the sky. Sometimes people call them "shooting stars." But they are not really stars. The real stars are *suns*, or balls of gases that glow by their own light. Our sun is only one of the billions of suns speeding through space.

The planets, their moons, the asteroids, and the meteors all circle our sun at great speed. They are the sun's family. They are held together by a terrific force called *gravity*. You have seen gravity at work. It is the "pull" that every object has for every other object—the larger the object and the more matter it has, the greater the pull. When you jump into the air, the earth's gravity tugs you back again. When you throw a ball, gravity finally makes it come down. When fruit drops from a tree, or when raindrops form in clouds, it is gravity that pulls them down to earth. Gravity pulls harder on some objects than on others, and so they are heavier. For the pull of the earth's gravity gives objects their weight. The earth's gravity is so strong that it makes meteors

A ball whirling on a string tugs away from the center of the circle it travels.

fall toward the earth, and it keeps our moon from flying off into space.

The sun's pull keeps the planets circling around it, instead of rushing off into space. There is something also that prevents the planets from falling into the sun. When objects move in a circle at great speed, they gather a force that makes them pull away from whatever is at the center of that circle. You can see how this works by whirling a ball on a string very fast around your head. The faster you whirl it, the harder the ball tugs away. In the same way, the speed of the planets creates enough force so that they tug away from the pull of the sun's gravity. The two kinds of pull balance each other, and each planet is kept moving in a set path around the sun.

Astronomers tell us that our sun *travels through space* at the great speed of over 720 miles a minute. It carries our earth and

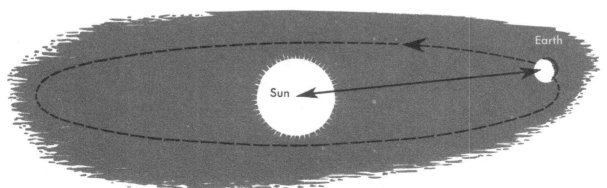

The earth, circling the sun, tugs away from it. Its tug is balanced by the pull of the sun's gravity, and so the earth is kept in its pathway.

the other planets with it. While the earth is speeding through space with the sun, it is also racing around the sun at more than 1,000 miles a minute. Even at this great speed, the earth takes a little more than 365 days to circle the sun completely. We call this length of time a year.

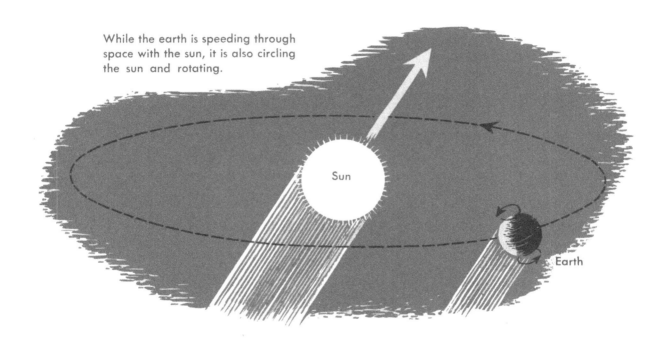

While the earth is speeding through space with the sun, it is also circling the sun and rotating.

While it is traveling through space with the sun and circling the sun, the earth is also spinning like a top. It makes one complete turn every 24 hours—our day. As the earth is 24,902 miles round at the equator, you can see that it spins, or "rotates," over 1,000 miles an hour there.

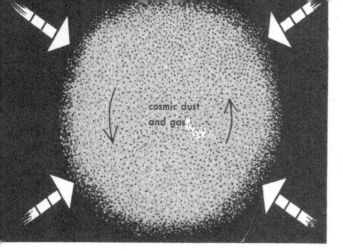

1. Some scientists think the planets may have been formed in the following way. A great cloud of gas and cosmic dust may have been spinning through space.

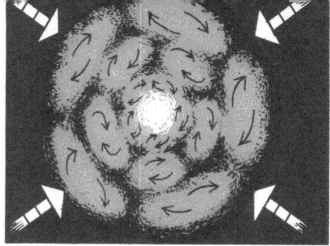

2. Gradually the cloud began to separate into large and small whirls.

how did it begin?

Scientists know that the earth travels through space, and that it is a planet of the sun. They know many other things about it, too, but they still do not agree on how the earth was made in the first place. Several explanations have been offered, but there is no real way of knowing whether any of them is right.

Some scientists have thought that the planets were once part of the sun. Possibly another star passed too close to the sun and pulled away some of its material; this material may then have become the planets, they say.

Other scientists have thought that our sun might once have been one of twin stars. They say the twin may have exploded, and its whirling material may have cooled into the planets.

Astronomers agree, though, that all stars are suns. And most of them agree that suns were formed from clouds of gas and the tiny particles of matter that float around in outer space. These particles are called "cosmic dust."

Some astronomers say that if stars were made from gas and

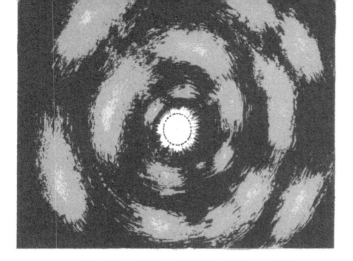
3. One of the whirls finally formed the sun.

4. Gradually solid particles in the smaller whirls pressed together to form the planets.

cosmic dust, the planets may have been, too. Something like this may have happened. Billions of years ago a great spinning cloud of gas and cosmic dust may have been traveling through space. Gradually parts of the cloud began to separate into large and small whirls. They collected more and more material to themselves. One of them spun around and around so rapidly that in time they became white hot. This large hot ball is our sun. Its gravity made it the center of smaller whirls of gas and dust. Gradually some of these smaller whirls joined together, and collected more and more material from the space around them. In each one the solid particles pressed tightly together until in time these whirls became our planets and their moons. Not all astronomers agree with this explanation. No one is sure.

Some astronomers say that in the beginning the earth may have been cold. But in any case, as it spun through the heavens and as its materials squeezed together it must have become very hot. Its particles slowly formed melted rock, very different from the rock we know today.

As the earth gradually cooled, its crust warped and great masses of rock pushed

Millions of years went by while the earth raced through the icy emptiness of space. Its material began to cool and harden and separate into heavier and lighter portions. The heavier material sank deep into the earth. Great masses of granite floated to the top. Granite is thought to be the lightest of the original rock. As the rock on the surface of the earth cooled, it started to form a thin crust. The earth did not cool quickly, however. Time and again the crust was broken. Underneath it hot rock pushed and shoved. Wherever there was a gap or weak place, the hot rock came pouring through to the earth's surface.

In places the crust was so strong that the rock pressing from underneath could not break through. But it could bend the surface upward. It pushed great masses of the outer rock up toward the sky. Hollows dipped between the high places.

8

upward. Meteorites from outer space fell and added to the earth's size.

Millions of more years went by. The crust warped and changed. And finally became solid. In the meantime, as the rocks cooled and pressed together, gases were squeezed out of cracks and holes. Many scientists think the earth may have been so small at first that its gravity would not hold the gases. They think, however, that many meteorites from outer space fell to the earth and added their material to its size. Finally it grew large enough to hold around it a layer of gas—its *atmosphere*. In the layer were nitrogen, oxygen, and the other gases that still make up our atmosphere—the air we breathe today.

At first the gases that escaped from the rock were very hot. When gas is hot, its tiny particles, called *molecules*, have great energy; they are very active. They move farther apart than before, and the gas becomes lighter and rises. This happened to the gas that escaped from the earth in that long-ago time. But as it rose far above the hot earth's crust, the gas cooled. In it was *water vapor*—water in the form of gas. As the molecules of water vapor cooled and lost their energy, they condensed; that is, they bunched together and formed tiny drops of water. The air became full of moisture, which covered the earth in a thick, dark blanket of clouds. The only light to reach the earth was from the flashing lightning, and flames that leaped above the great gaps in the crust. Rain may have fallen from the thick clouds, but for a long time the earth must have been so hot that raindrops nearing the rocks were changed again to water vapor.

Finally, the earth cooled enough so that rain could fall on its surface. Then it rained and rained and rained—no one can say for how long. The rain probably lasted for years. The earth was nearly drowned. The drops that fell on the high places streamed down into the hollows. So the oceans began to form. Now the earth had air, water, and bare solid rock.

Finally the earth cooled enough so that torrents of rain fell, and the oceans formed.

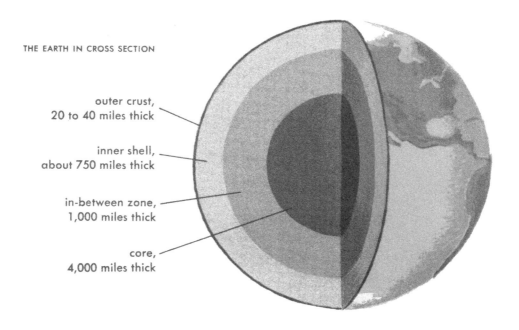

THE EARTH IN CROSS SECTION

outer crust, 20 to 40 miles thick

inner shell, about 750 miles thick

in-between zone, 1,000 miles thick

core, 4,000 miles thick

what is the earth made of?

Geologists think the rocky sphere of the earth is made up of the following main parts:

1. The outer crust is probably from 20 to 40 miles thick. Not all its rocks are of the same kind or of the same weight. The heavier ones sank to form the basins of our oceans. The lighter blocks of granite rose to make the beginnings of our continents. Underneath the granite, and mixed with it, is a heavy black rock called basalt. Some geologists call this basalt the "basement of the continents and oceans."

2. An inner shell, or *mantle*, is about 750 miles thick. It is a mixture of very heavy rock.

3. An in-between zone, about 1,000 miles thick, is made of rock and iron.

4. The *core,* or center, of the earth is a sphere, probably about 4,000 miles through. There is still much to be learned about it. It is thought to be a mixture of iron and nickel.

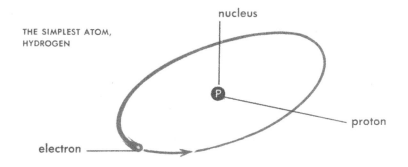

Scientists in various fields are working to find out more about the inside of the earth. Astronomers try to learn about it by studying the other planets and the meteorites. Geologists study the earth itself and its rocks. Chemists and scientists called physicists study the earth's materials to see what they are made of and how they are put together. The chemists and physicists have already given us the answers to this question: why are there so many different kinds of rock in the earth?

All the materials in the world can be divided into three kinds: *solids, liquids*, and *gases*. Rock has length, width, and thickness. It is a solid. Water runs. It is a liquid. Air has no fixed shape, and it can spread out, or take up more space, without any limit. It is a gas.

Solids, liquids, and gases are all made of tiny particles called molecules. They are so small that they cannot even be seen with a microscope. For a long time scientists thought molecules were the smallest particles there were. After many experiments, however, they found that molecules are made of even smaller particles called atoms. Now the scientists have proved that even the atoms are made of something smaller.

Of course, no one has ever seen an atom, but complicated experiments have shown that it is somehow made of electricity. The nearest we can come to understanding how an atom might look is to compare it with the sun and its planets. The simplest atom of all is that of the gas hydrogen. At its center, or nucleus, is a particle called a *proton*. This has a charge of electricity that is positive—it has a plus value. Outside the nucleus, and circling around it, is another particle, an *electron*. This has a charge of electricity that is negative—it has a minus value. The positive charge of the proton and the negative charge of the electron attract each other and so the atom is held together. More complicated atoms than hydrogen have particles called *neutrons*, which are like protons, but have no electrical charge.

Protons, neutrons, and electrons are the main building blocks of the atoms. Each single positive proton can attract and hold a single negative electron. Because of this the number of protons in an atom decides its number of electrons, and what kind of atom it is. A nucleus with one proton attracts one electron, and makes an atom of the gas hydrogen. A nucleus with two protons attract two electrons and makes something different: an atom of the gas helium. An atom with three protons will, in general, have three electrons. This makes an atom of lithium, a soft, silver-white

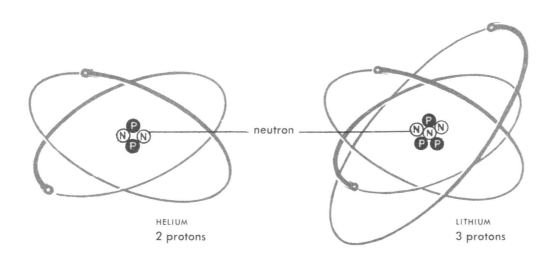

HELIUM
2 protons

neutron

LITHIUM
3 protons

material. Of course, each different kind of atom forms a different kind of molecule. And each different kind of molecule makes a different material.

Everything around us is made of various combinations from among the earth's simplest materials, called *elements*. An element is so simple that it cannot be separated by any chemical method into things simpler than itself. There are over 90 natural elements and each is given an atomic number according to the number of protons it has in its nucleus—that is, from 1 to over 90.

Certain elements have atoms that lose some of their electrons easily. When this happens, the positive protons and the negative electrons in the atom no longer balance. The atom has more protons than electrons and so it becomes positively charged.

Other kinds of atoms gain electrons easily. Then they have more negative electrons than they have positive protons and they become negatively charged.

Negative and positive atoms attract each other. They join together into a new kind of molecule: a combination, or *compound*, of the original elements. When this happens, the molecules make a new substance. Sodium, a soft silvery element, parts with one of its electrons easily. Chlorine, a green poisonous gas, will not part with an electron, but it will quickly take one from sodium. And what a surprise we have! When these elements combine, they become a compound we use every day: our ordinary table salt.

In this way, by gaining and losing electrons, and also, sometimes, by sharing electrons, the elements join together in various combinations, or compounds, to form other substances.

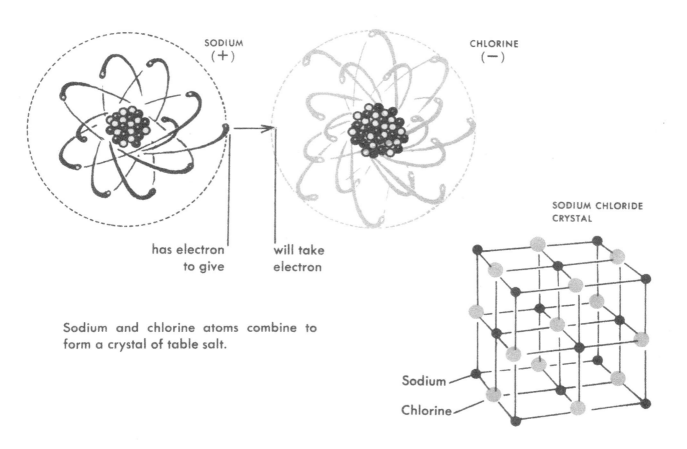

Sodium and chlorine atoms combine to form a crystal of table salt.

The earth's crust is made up almost entirely of only eight elements in different combinations. While the crust was still hot, the atoms of these elements traveled about easily and joined with one another to make various new substances. How they joined and what they made depended on many things. If the material through which they moved was completely melted they could travel farther than they could if it was partly cooled and thick as tar. If there was a great deal of water or if there were many gases, that made a difference too.

The natural substances that the elements formed in the crust are called *minerals*. Rocks are either simple materials or, more often, mixtures of minerals.

Waves beat against the shore and wear the rocks away.

the rock families

Geologists put rocks into three large families.

The first family is *igneous* or "fire" rock. (Igneous comes from the Latin word for fire.) Igneous rock forms from hot melted rock material, as it cools and becomes solid.

There are many different kinds of igneous rock, because of the various combinations of elements and the conditions under which they joined. If the hot materials cooled quickly, they made one kind of rock. If they cooled slowly, they made another. If they cooled deep inside the earth, and very, very slowly, they formed a still different kind of rock.

Probably the most important igneous rocks are granite and basalt, which make up large parts of the earth's crust. Almost everyone has seen granite, and noticed the flecks of the minerals that came together to make it.

Almost half the material in igneous rocks is the mineral feldspar. Feldspar gives rock many of its white, pink, red, and gray colors.

Nearly one-fourth of the remaining material in igneous rocks is the mineral quartz. This appears in many colors, all of which have a somewhat glassy look. Other minerals appear in smaller quantities in igneous rocks.

Igneous rocks were the earth's first ones. Almost at once they began to change. No sooner had the atmosphere been created than it began eating away at the earth's crust. Certain gases of the air attacked certain minerals in the rocks. This made new substances and helped wear the rocks away.

When at last rain poured down upon the earth the water joined with minerals to make acids which ate away more of the rocks. Pieces of them split off. Tons of broken material fell from steep cliffs into the ocean. Here the waves picked up the pieces and pounded them against the cliffs until they were worn into tiny bits of sand, mud, and clay.

You can see for yourself how the rocks are still wearing out, day by day. At the seashore, piles of broken pieces lie at the foot of the cliffs. The waves wash back and forth, scouring the rocks smooth. When there are storms, the breakers hurl pieces of rock against the cliffs, and carve out giant holes and caves. The sand and pebbles you find on the beach were once part of the rocks.

THE ROCKS WEAR AWAY

Rocks in river beds are gradually ground to pieces.

Ice breaks the rocks apart.

Roots pry the rocks apart. →

The rocks on the mountains are forever changing, too. The sun heats them in the daytime, and causes them to expand, or swell. At night, when the air grows cold, they contract, or shrink. As this goes on over and over again, the rocks crack. Rainwater creeps into the cracks and freezes. The ice pushes hard, just as it pushes the cap up from a bottle of milk on a winter morning. All this is called "weathering." Finally, the rocks fall apart, and pieces tumble down the mountain, breaking more and more as they fall. Piles of their shattered bits spread out like fans from the base of the cliffs. If you look, you may find smaller rockslides like this near your home.

Notice also how plants wear away the rock. The roots of trees and other growing things work their way into cracks. As the roots become larger, they pry the rock apart. Rotting leaves join with rainwater to make acids that decay the rock. Day by day it falls apart. Rotted bits of bark, leaves, and other pieces of once-living plants and animals join with the rock bits to make the soil that covers much of the earth's crust.

Long, long ago, when the rains poured down upon mountains, the rushing water picked up pieces of broken rock and carried them along. The water ran where there were natural troughs in the earth's crust, and made creeks and rivers. All along the river bottoms, the rock chunks banged and scraped until even the hard bits of quartz and feldspar were ground into tiny grains of sand. At last, tons and tons of sand, pebbles, mud, and clay reached the lakes or shallow seas and slowly sank to the bottom. Flood after flood dropped more of this material, called *sediment*.

One layer of sediment after another settled on the floor of the

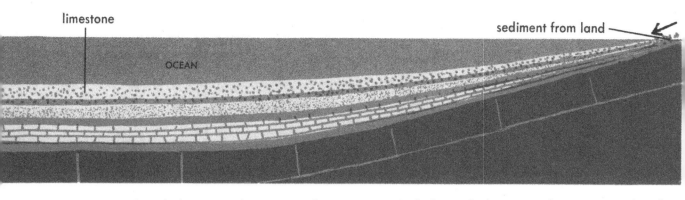

Layers of rock are formed from sediment on the ocean floor.

quiet lakes and seas. The great weight of the top layers and of the water pressed those underneath until they became hard. The minerals in the water helped cement the tiny particles together, and the sediment became rock. This so-called "sedimentary" rock, formed in layers, is the second big family of rocks. Sedimentary rock is still being made today. Wind, as well as water, sometimes piles up sediment that later becomes rock.

You probably know some of the common sedimentary rocks. Sandstone, which looks and feels like sand, is made mostly from quartz sand. Limestone is made from line and the shells of dead water animals and from lime in the water. Shale is made from mud and clay.

The rocks of the third family are the *metamorphic*, or "made over," rocks. They are made from igneous and sedimentary rocks.

From the time gases became water and hot rock cooled, the earth has been constantly changing on the outside. The inside is forever changing, too. Far below the earth's surface, solid rock melted until it was soft, like dough. The hot doughy rock is called

Heat and pressure change sedimentary rock to metamorphic rock.

"magma" when it is inside the earth. When it pours through cracks in the surface, it is called "lava."

No one is certain why the below-surface rock first started to melt. Possibly the great pressure from the rocks above, and the earth's gravity pulling from below, had much to do with it. Moreover, when certain gases change into other substances, heat is produced.

Once the rock was heated enough to be doughy, it melted the solid rock around it. Then it began to move. Hot doughy rock is bulkier and lighter than hard rock, so it slowly oozed upward. It squeezed and pushed. Water, gases, and various acids helped it. Slowly it changed the minerals in the rocks it pressed against and heated. They were made over into metamorphic rocks.

Igneous rocks like granite were made over into gneiss (pronounced "nice"). Limestone was changed to marble. Shale was changed to slate or schist (pronounced "shist").

Heat and steady squeezing from above, without any other motion, can make over rocks, too. A layer of shale, pressed long and hard, and mixed with acids and gases, can become slate.

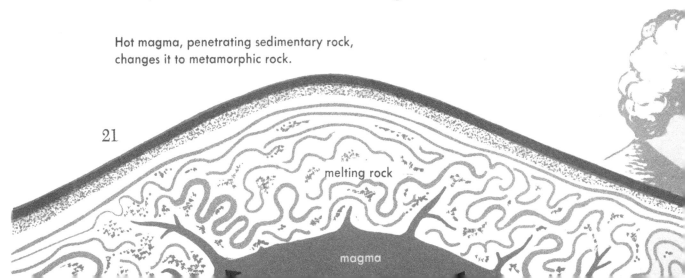

Hot magma, penetrating sedimentary rock, changes it to metamorphic rock.

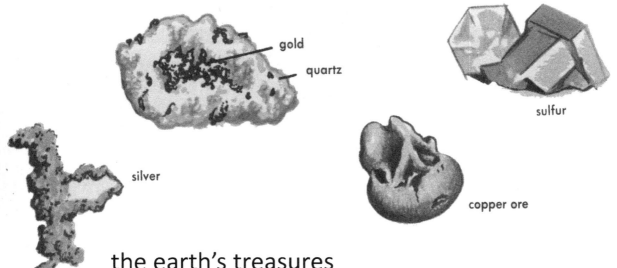

the earth's treasures

All this tearing down and pressing and melting of the rocks has given our earth great wealth.

In some places, ancient forests were covered with water, then with layers of clay and mud. As the layers were pressed into rock, the dead trees that were caught in them gradually changed to coal.

In other places animals and plants decayed in shallow water and were covered with sediments which later became rock. Oil and gas were squeezed from the dead plants and animals, and became the petroleum and natural gas we prize so highly today.

Iron minerals settled in layers on the bottom of ancient seas and were covered with sedimentary rock. Today they make our riches iron mines.

While the hot rocks were melting and changing, large amounts of valuable metals often joined with other elements to make ores. By mining and melting the ores, we can separate the rich metals from the rock.

Gold is found in igneous rocks in some places. It is also found in some sedimentary rock.

Zinc, copper, tin, mercury, aluminum, and uranium are only a few of the other metals that form valuable ore deposits.

the bending crust

But, you may be wondering, how do we know about sedimentary rock if it is formed under water? And how do we know about metamorphic rock if it is made over deep inside the earth?

We know about them because we can see them. In the sides of cliffs we can sometimes see sedimentary rocks, one layer above another. Even high in the mountains there are sedimentary rocks with the shells of sea animals in them. They were formed in the bottom of the ocean. And in quarries we find slate and marble, both metamorphic rocks.

But this is even more of a mystery! If the earth is always wearing down, how did rock from the ocean ever get up into the mountains?

The answer is this. The earth is always wearing down, but it is forever changing in other ways, too. The crust of the earth can

Layers of sedimentary rock can often be seen in cliffs. These layers occur at Muckross Head, Donegal County, Ireland.

bend. Parts of it have moved upward or downward or sideways many times since it was first formed. During thousands of years sedimentary rocks from beneath the sea have been slowly lifted upward in several places. They end as high platforms or *plateaus*. Often their layers still lie flat and in the same order as when the ocean washed over them long ago.

A folded towel, pushed inward toward its center, illustrates how rock layers fold.

Under some conditions rock layers can be folded. This happens when there is strong pushing deep inside the earth—from the gases and big masses of magma that press hard, and from the layers of sediment that get heavier and heavier as they grow thicker. Mountains as high as the Rockies have been made by this strong pushing.

Take a heavy towel. Fold it back and forth several times lengthwise, and lay it, still folded, on a table. Put one hand on each end of the towel and push it slowly toward the summer, until there are

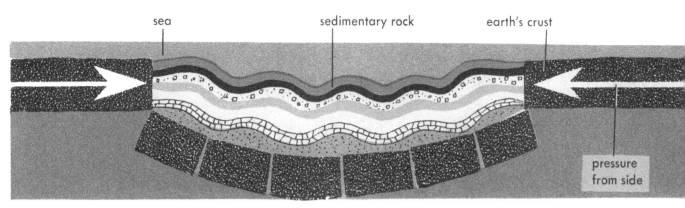

Layers of sedimentary rock beneath the sea start to fold when pushed from either side.

several humps. These represent hills or mountains. The layers of toweling represent layers of sedimentary rock, such as limestone or sandstone, which used to lie flat, one above another, on the bottom of the sea.

This is what happens. As the earth is worn down, layer upon layer of sediment is washed into a shallow sea. The sediments grow deeper and heavier. The bottom of the sea keeps sinking under their weight, until finally the layers are several miles thick. Then gradually there is pressure from the sides. The flat layers of sedimentary rock on the sea bottom are pressed into wavy folds, just as your toweling was. Finally, the bottom of the sea is slowly raised, until the folded rocks become hills or mountains.

Rocks fold in various ways. Sometimes the pushing is greater on one side than on the other, and one of the humps is pushed

When the bottom of the sea is slowly raised, the folded rocks form mountains.

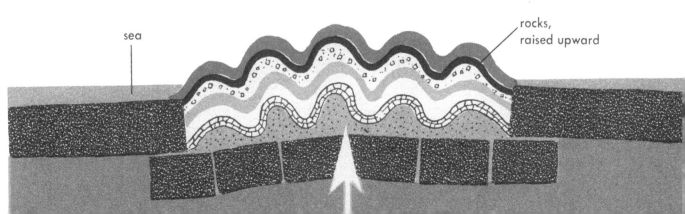

The Rocky Mountains at Banff, Canada, show tilted strata.

right over the top of another. This makes the layers upside down. Once in a while the pressure is only on one side or under one end of an area. Then the rocks are pushed up, high on end, and tilted, usually into steep cliffs. Rocks tilt in many ways, and make several kinds of mountains. The Sierra Nevada Mountains in the West are a tilted range.

If you have ever tried to make a bow and arrows, you know how tough willowy wood will bend and bend if it is pushed slowly. Some other kinds of wood are brittle, and will snap if they are pressed too hard. In a similar way some rocks bend more easily than others. In order that rocks may bend and fold, the pressure must be very, very slow. Perhaps the rocks will bend only a few inches in hundreds and hundreds of years.

Sometimes the pressure is too great or too sudden. Sometimes the rocks are weak in spots. Then they snap and crack and break, instead of folding.

Often the great mass of rocks on one side of a crack slips up or down or sideways past the rock on the other side, and in a different direction. This is called a *fault*.

When rock moves up or down a fault very suddenly, it is as if the earth had been struck a heavy blow. It trembles, and the shaking motion, called a *shock wave*, goes out through the rocks in ripples, similar to what happens when you drop a pebble into a quiet pool. This sudden movement of rock is an *earthquake*. Although earthquake ripples in the rocks may be very small themselves, they cause enough trembling to make walls crack and buildings crumble near the center of the earthquake.

After an earthquake there are three kinds of shock waves. Two of them take a shortcut through the earth, and the other one follows the surface. The first wave that passes through the earth is much faster than the second one. The surface wave is the slowest of all, but it is large and does a great deal of damage. A record of all these waves can be made by an instrument called a *seismograph*. This has a heavy pendulum, balanced so that it stays still

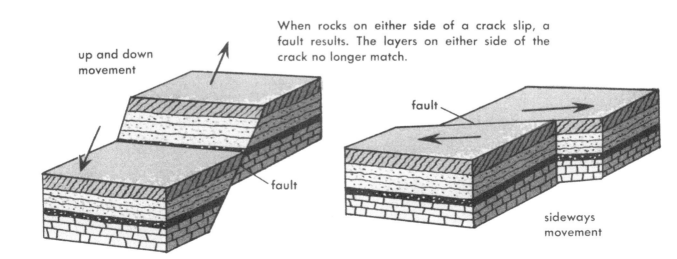

When rocks on either side of a crack slip, a fault results. The layers on either side of the crack no longer match.

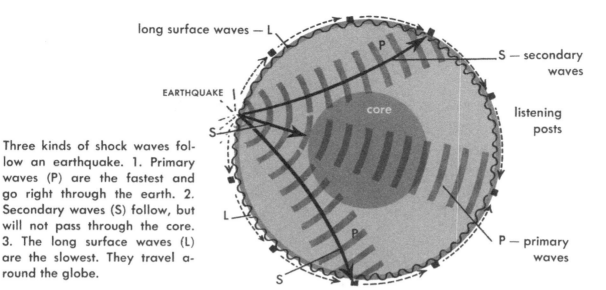

Three kinds of shock waves follow an earthquake. 1. Primary waves (P) are the fastest and go right through the earth. 2. Secondary waves (S) follow, but will not pass through the core. 3. The long surface waves (L) are the slowest. They travel around the globe.

if the earth moves beneath it. Fastened to the pendulum is a stiff, slender rod which holds a pen. The pen's point, in turn, rests upon a piece of paper. During an earthquake, the pendulum, the rod, and the pen stay still, but the paper moves with the earth and the pen marks these movements.

In the very newest seismographs a pen is not used. Instead, the earth's shaking is photographed. By studying how these shock waves act as they go through the earth, scientists have learned much about its inner parts.

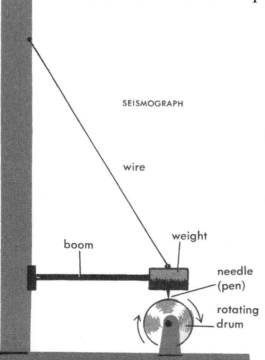

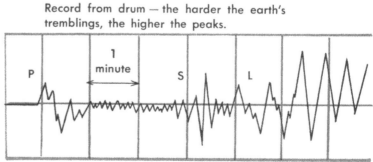

Record from drum — the harder the earth's tremblings, the higher the peaks.

A seismograph's needle records the earth's shaking on paper attached to a rotating drum.

volcanoes

Some kinds of mountains are made by hot underground magmas. As these magmas force their way slowly upward, they push like giants on the earth's crust. Sometimes they reach rocks with water in them, and steam forms. This, with the gas in the magmas, presses even harder on the crust. And sometimes, as the magmas

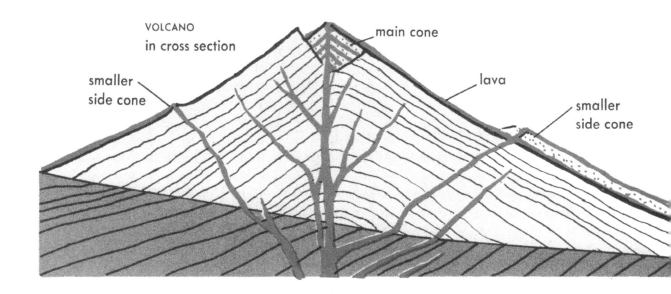

approach the surface, they reach a weak spot. Then the gas and steam burst out through a hole in the earth with a great explosion. Rock, cinders, ashes, and the lava are blown upward. Tremendous earthquakes follow. Hot lava often pours out through cracks in the crust near the opening. This disturbance is a *volcano.*

As its cinders, ashes, and dust fall to the earth, they make a cone around the volcano's opening. Each outburst of the volcano makes the cone larger. Mount Pelee, on the Island of Martinique in

The volcano Paricutin and its cone of lava and ash.

the West Indies, and Mount Vesuvius, in Italy, were built in this way. Much more recently, in 1943, an opening appeared in a cornfield in Mexico and hot lava poured from it. Two villages were destroyed, as lava and ashes spread out over the countryside. This volcano, Paricutin, grew to be several hundred feet high before it stopped erupting in 1952.

There are about 400 active volcanoes on the earth today. They seem to occur only in certain areas and are not scattered widely over the whole world. Many of them form what geologists call a "ring of fire" around the Pacific Ocean. Possibly the earth's crust in this area is weak.

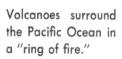

Volcanoes surround the Pacific Ocean in a "ring of fire."

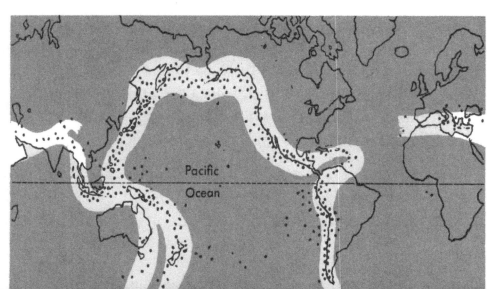

hot rock underground

Volcanoes become less active as they grow older, until finally they die down altogether and become cold. But there is a time during which there is still hot rock underground. Then, instead of lava, hot gas comes from the volcano's opening.

THE VALLEY OF TEN THOUSAND SMOKES, ALASKA

Sometimes water reaches the hot underground rock from the earth's surface, and is changed to steam. Then gas and steam may escape through cracks in the ground in jets that are called *fumaroles*. Near the Alaskan volcano Katmai there is a valley hazy with so many fumaroles that it is called the Valley of Ten Thousand Smokes—and it looks just like its name!

If there happens to be water running underground near the hot rock, this may possibly become heated and may bubble from the ground as a *hot spring*.

Geysers are midway between hot springs and fumaroles. They give off both hot water and steam. This is how a geyser works. Water from the earth's surface seeps down through cracks until it reaches the hot rocks. As it heats, it rises through an opening or tube. And as the water gets hotter and hotter, steam forms under it, and presses. Finally the steam presses so hard that the water is blown through the opening and high into the air, like a fountain. This is a geyser.

As the water blows out of the tube, the pressure becomes less and the geyser spouts lower and lower, until it finally stops. The water runs back into the ground and starts to heat all over again. Some geysers blow off so regularly that people know exactly when to watch for them. Old Faithful geyser in Yellowstone National Park spouts about every 65 minutes. As time goes on and the

Underground water heated by magma sometimes bubbles from the ground in a hot spring.

OLD FAITHFUL GEYSER, YELLOWSTONE NATIONAL PARK

Underground cross section of a geyser

underground rocks grow cooler, the geyser will go off less often. Finally Old Faithful will stop spouting altogether, and will become a hot spring.

More often than not, magma forces its way into the surrounding rock but cools before it reaches the surface of the earth. The rock it enters may be of a different kind entirely.

Sometimes magmas move upward as one huge mass. They melt and cook hundreds and hundreds of square miles of the surrounding rock, but do not build up any great pressure. These enormous masses are called *batholiths*. They cool slowly underground, and a great variety of minerals have time to form. Ores such as copper and silver may be found in them.

When magma leaves the main mass and creeps *upward* through cracks in the surrounding rock, it forms *dikes*. When it spreads *sideways*, between layers of rock, it forms *sills*. Dikes and sills are small as compared with batholiths, but some of them have covered many square miles of the earth.

Sometimes, when sills start to form, more and more magma collects and hardens. Then it makes huge domes, called *laccoliths*, which force the underlying layers of rock to hump upward as hills or mountains. The Henry Mountains of Utah are laccoliths.

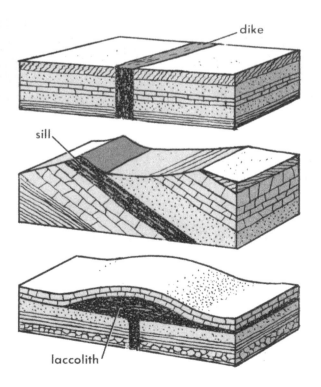

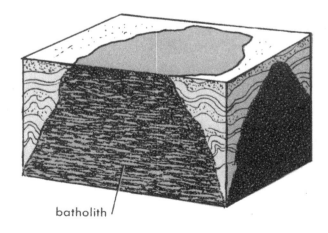

Hot magma forces its way up through cracks in the surrounding rock to form dikes, sills, laccoliths, and batholiths.

Pike's Peak, Colorado, a batholith which was originally formed underground.

The rocks that form deep in the ground may yet appear as bare mountain peaks. As the land rises in places to make the hills, soft overlying rock may slowly wear away in wind and weather. Then the top of a batholith which was once deep underground, may become the crest of a mountain. Pike's Peak, over 14,000 feet high, in Colorado, is a bare batholith top. But the bases of batholiths go so deep into the earth that no one has ever seen one.

Even in your own neighborhood you are likely to find sills and dikes in the rocks. Watch for streaks and stripes that look entirely different from the rocks through which they pass. Chances are that at some ancient time those streaks were hot magma working its way upward or sideways through small cracks in older rocks.

down come the mountains

As fast as the earth is built up, it is worn down, and so the underlying rocks are bared to tell us their story. The wearing down forces of the earth fight against the great mountain-and-rock building forces. Everywhere you go you can see their work. The piles of broken rock at the foot of a cliff, the muddiness of a flooded brook, the little whirl of dust that dances down the street on a windy day, all are reminders that the land is being slowly torn apart. Building up, wearing down, building up, wearing down—that is the earth's history. And that is what gives our world the curves of its hills, the sweep of its rivers, the shelter of its valleys, the sudden edges of its shorelines—all its endless variety.

Since the rocks were first formed, their two great destroyers have been air and water, which together can enter the smallest crack in the earth's crust.

In many places moving air, or wind, has built and destroyed land without the help of water. In dry stony areas where strong winds blow, tiny grains of sand and small pebbles are picked up and whirled against the rocks. There they grind and rub just as sandpaper would, to carve many strange shapes.

There are deserts in parts of the earth where there is little or no rain. Some of them are sandy. Very few plants grow to help anchor the soil, and the wind blows the sand into ripples and the little hills called dunes. Near lakes and ocean shores tons of sand also pile into dunes. A tiny bush, a rock, or even a tin can often halts the wind enough so that it drops some of its load of soil.

The Badlands of South Dakota have been worn away by wind and water.

A natural bridge in Monument Valley, Arizona and Utah, has been carved by blowing sand.

Rain wash, wind, and sand have sculptured these strange pillars in Monument Valley, Arizona and Utah.

Dunes may sometimes bury trees and houses.

Once a mound is started, more and more sand is added, until the mound grows into a giant dune. Some such hills have grown three or four hundred feet high, and have buried farms, cottages, and trees along the shore. The great sand dunes of Colorado and those along the shores of Lake Michigan show what the wind can do.

a river's story

But running water is the greatest natural carver of the earth. Every heavy rain makes new little troughs or gullies, and water flows down them to creeks and rivers.

If the land is high, the slope to the sea is steep. Then the water runs quickly and sweeps heavy rocks along with it. With their help the river slices sharply downward, and makes a V-shaped valley.

Of course the river cuts into soft rocks much more quickly than

A young river valley is V-shaped.

it does into hard ones. So there is likely to be a waterfall where the riverbed changes from hard rock to soft rock that is easily worn away. Or sometimes a fault in a rock makes a cliff, over which the river drops to make a waterfall.

During thousands and thousands of years the land is worn lower and lower, and the river runs more and more slowly. Finally, it can no longer carry heavy rocks, and so it stops cutting downward quickly. Now it carries mostly mud and sand. The sides of the valley have worn away so that it no longer is V-shaped, but has grown wider. The river wanders back and forth in gentle curves, and makes the valley wider still.

An old river valley is wide, and its river wanders in curves.

CROSS SECTION OF A DELTA, showing how it is formed.

Now that the stream runs more slowly, tons of its load of mud and sand drop to the river bottom. This makes the water higher. Then the river overflows after every long rain, and causes a flood. As each flood ends, the water leaves a layer of mud and clay on the flat valley floor. Gradually a *flood plain* has been formed. Millions of people live on flood plains like the Nile and Mississippi River valleys because the soil is rich. They must often fight to prevent floods, however.

When the river enters the ocean, its running water slows down. Then the stream drops the rest of its load of mud and sand. Some of the load gradually builds up into a fanlike piece of land spreading out from the river's mouth. This land has somewhat the same triangle shape as the Greek letter *delta* (▲). So it has been given the same name: delta. As more and more material comes down the river, the delta grows in length. It forms a swampy flatland that is easily flooded but is very rich for farming. Some of the most ancient civilized peoples lived on the deltas of the Nile and the Tigris and Euphrates rivers.

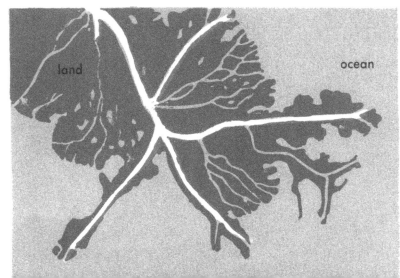

A delta spreads out from the river's mouth.

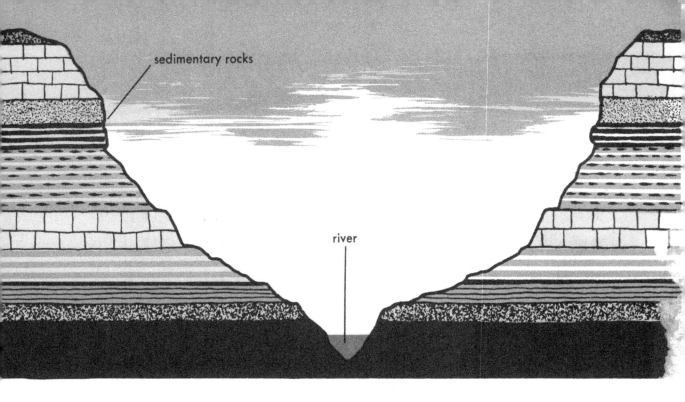

CROSS SECTION OF A CANYON

the earth's history book

Sometimes streams start where there is plenty of water, but on their way to the ocean they run over a high, dry rocky plateau. Then the river they form cuts sharply downward because the land is so high. The valley sides cannot be worn away quickly by rain because the climate is dry. Deep, steep-sided valleys like this are called *canyons.* Canyons in soft sedimentary rock are often sliced hundreds of feet downward by the river. On their sides can be seen the many-colored rock layers that would be underground in other places. Usually the layers lie almost flat, much as they were formed: the oldest layer on the bottom, the next oldest layer just above it, and so on, up to the youngest layer, which lies at the very top. They are like a stack of books on a table. The one at the bottom was put down first, the second from the bottom was put down next, the top one was put down last. If the layers are read in order, from the bottom to the top, they can reveal much about the earth's history at that place.

Back in 1869, a man named Major John Wesley Powell realized that one of the best records of the earth's history anywhere lay in the Grand Canyon of the Colorado River. At that time no one knew very much about the canyon, except that it was probably a mile deep in some spots, over ten miles wide in others, and hundreds of miles long. No one had ever explored it. There were frightening tales that it ran underground for over a hundred miles, that it had high waterfalls, and that anyone who tried to go through it in a boat would never come out alive.

But Powell was determined to explore it. He had studied rivers, and he saw that the Colorado was an unusually muddy one. By now, he reasoned, a river as muddy as that, which had made so deep a valley, surely must have worn away any waterfalls there were. He expected bad rapids only. He ordered four boats especially built to withstand the sledge-hammer pounding of the rocks and rapids. On May 24, 1869, he and nine companions started down the Green and Colorado Rivers from southern Wyoming. The friends who saw the explorers off expected never to see them again, and for months nothing was heard from them. Again and again they were reported dead.

It is true that they had many narrow escapes. The rapids were terrible, and the men spent almost as much time in the water as they did in the boats. Often they stood on the banks and let their boats downstream by rope, hand over hand, in places too rough to navigate. Within a month one of the boats was wrecked and the men barely rescued. Soon after, the group lost almost all its cooking equipment.

Later the towering canyon walls closed in so that the men could no longer stand on the banks and let their boats downstream. Then they had to ride the roaring, frightening rapids. Their food supplies were so often soaked with water that for a large part of the way there was nothing to eat but moldy flour and rancid bacon. Then they began a race with time, to get through the canyon before their food was gone. Three men gave up, and made their way out of the canyon on foot. The others kept on, and on August 29, they suddenly left the canyon and came out into a quiet green valley.

What they had seen in the canyon was far beyond anything they had imagined. Here was a record of the earth that Powell figured must go back millions of years. Reading it, he understood the earth's movements and the work of rivers as he never had before. At the foot of the canyon were the worn-down roots of ancient mountains. They were filled with volcanic rocks and dikes. At some distant time the sea had flooded over them and had laid down heavy layers of sedimentary rock. After that the earth must have been uplifted, for Powell could see where the sedimentary

rock had been worn down by weathering. Then once more the sea had moved in, and more sedimentary rock had been laid down. Once more the earth must have been raised, for wind and rain had eaten away part of that rock layer, too.

From what he saw on the many-colored sides of the canyon, Powell figured that the sea had covered the area three times; the land had lifted three times; and volcanoes had poured out lava three times. He could see that at the present time the whole region had been slowly lifted to a high plateau. During the thousands of years of the lands uplifting, the river had cut down, now slowly, then quickly, to make the canyon. Wind and water had carved away some of the canyon's softer stone and had left the

harder rock in natural bridges, table rocks, buttes (small, flat-topped hills), and mesas (large, flat-topped hills).

Today tourists can visit Grand Canyon National Park, in Arizona, and see the canyon, a mile deep in places, and eighteen miles wide at some points. It is one of the most impressive sites in the world.

LOOKING OUT OVER THE GRAND CANYON, ARIZONA

ground water

Of all the rain that falls only about a third runs off in streams and rivers. Most of the rest goes underground. It soaks into the soil, and follows joints and breaks in the rocks. From there it works its way into every little crack and pore it can reach. It spreads slowly underground much as ink soaks into blotting paper.

This ground water, as it is called, soaks the rocks and soils in a

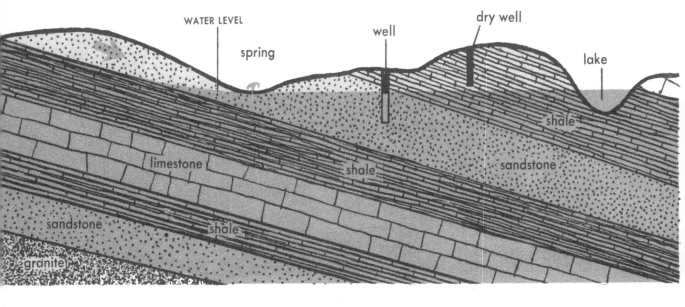

band fairly near the surface of the earth. The top of this band is known as the *water level* or *water table*. The water table goes up or down, depending on the rainfall. In a dry season it goes down. In a wet season it goes up. Wells driven or dug into the earth are kept supplied by ground water. But they must go deep enough to reach below the water table even in a dry season, or they will sometimes fail.

Much of the ground water comes to the surface again. Sometimes the water table is high enough to break out of the earth in places. Then ground water, trickling through rocks, higher up a mountain or hillside, may flow out of a crack or opening at the water table, as a spring. Or ground water may seep into low, spongy soil to make marshes and swamps, which are always damp. Springs or slow oozings from the water table sometimes fill hollows in the land and make ponds and lakes. Of course, a lake or pond may also be made by a natural dam across a stream. Hills may form one; beavers may build one; or lava may harden across a river in a solid flow, and make a dam.

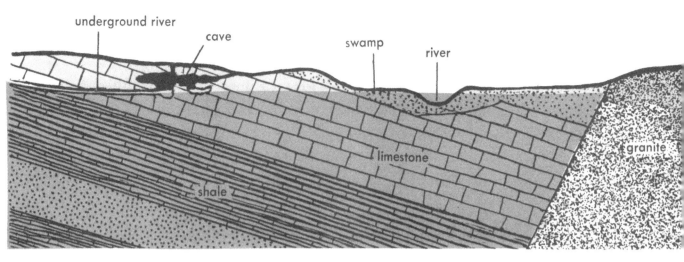

A great deal of ground water goes back into the atmosphere as water vapor. There is always water in the ground, however, and there it does surprising things. Water joins with various minerals, and works all kinds of changes. It dissolves minerals out of the rock, picks them up, carries them along, and then drops them somewhere else. Sometimes it helps change them to different minerals altogether.

Some rocks, such as limestone, gypsum, and salt, are dissolved quite easily. When water keeps moving through breaks in these rocks, it may gradually eat them away until the openings become *caves*.

Water often forms caves in limestone.

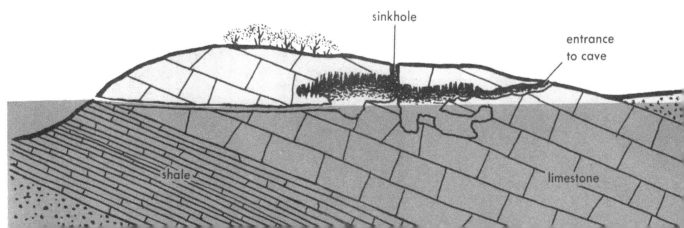

Once the caves are made, water that enters them usually drips down and makes a stream along the floor. Sometimes rivers on the surface of the ground disappear suddenly through openings in the earth. The streams run through caves for some distance, then suddenly come out of the earth again. These are known as *lost rivers*.

Water may dissolve so much rock that the cave's roof falls in and makes a hole into the cave from above. This is called a *sink hole*. Sometimes almost all the roof is worn away except a narrow piece above the cave. This makes a *natural bridge*.

Water may gather so slowly on a cave's ceiling that it evaporates—turns it to water vapor—before it can fall to the floor. Then, with the liquid water gone, the mineral that was in it is left clinging to the ceiling. More and more material joins it as water keeps gathering and evaporating. Long, lacy mineral "icicles" gradually

NATURAL BRIDGE, VIRGINIA

CARLSBAD CAVERNS, NEW MEXICO

stalactites

stalagmites

A SINKHOLE

PETRIFIED FOREST, ARIZONA

form. They hang from the ceiling in odd shapes call *stalactites*. At other times, the water may drip to the cave's floor, but so slowly that it evaporates there. Then its minerals pile up in shapes called *stalagmites*. Ground water, slowly dripping, can change caves into strange and beautiful places.

Water that spreads through the soil and rock sometimes reaches a seashell or what is left of a piece of wood or some other once-living thing. Then the water may take away the material of the shell or wood, particle by particle, and instead may leave minerals such as silica or iron. An exact mineral copy—a petrified copy—of the original shell or wood is formed. Often this is in rich colors, and is very beautiful. There are numerous whole petrified forests in the western United States, among them Petrified Forest National Monument, in Arizona.

glaciers

Besides water underground, water on the surface of the earth in lakes and rivers and oceans, and water in the air as water vapor, there is water as ice.

In cold regions of heavy snow—the Arctic and Antarctic, and high on mountains—more snow falls in winter than melts in summer. Suppose that six feet of snow fall and only five feet melt. If this goes on year after year, the snow piles up. After a long time it forms a thick mass, even in summer. This mass may make a *glacier*.

In the hot summer sun, a little of the top snow melts, and the water trickles downward. At night the air is colder, and the melted water freezes. As new layers of snow are added on top, the pressure on the older, lower layers becomes very great. Gradually the snow is squeezed and frozen into a huge mass of ice—a glacier, sometimes hundreds of feet thick.

Of course, all this snow and ice is very heavy. The bottom of the glacier starts moving to "get out from under" the great weight of the top. Now the glacier becomes a giant tongue of ice, slowly "flowing," often down a valley. The glacier does not flow quickly, like a river. Possibly it goes only a few inches a day—so slowly that no one can see it moving.

Louis Agassiz, a young Swiss, lived on a glacier to study it. He was the first to prove that glaciers really do flow. He drove a straight line of stakes deep into the glacier's ice and into the valley walls on either side. Sure enough, after a time, the stakes on the glacier moved forward, while those on the valley walls stayed exactly where they were.

As a glacier presses down, its ice freezes to the rocks and soil beneath it. Dirt and more rocks roll down on the top from the valley sides, and are frozen there. The river of ice inches forward, carrying all this material with it.

The rocks on the underside, next to the earth, become the glacier's tools. They rasp and file away at the earth, and break away more rocks to join the ice. They scrape and scour rough places smooth. The walls of V-shaped valleys are dug away, and the valleys become U-shaped, as if a giant thumb had scooped them out. Hollows are dug, which may later become the basins of lakes.

Finally, the glacier reaches a warmer region. There it starts to melt. Streams run from underneath it, and carry away mud and gravel and rocks. As the ice disappears, huge boulders that were frozen in it are left stranded in the ground. At the edge of the glacier, ridges called *moraines* are made of soil and rock that drop from melting ice. They may become dams for future lakes. Long rounded hills called *drumlins* are also formed. The whole surface of the land is changed by material from the melting glacier.

Boulder, stranded in a field

U-shaped valley

Land that has been under a glacier is easy to recognize, once a person knows what to look for. After Louis Agassiz had studied glaciers, he realized that much of Canada and the northern United States and northern Europe must once have been covered by them. He could see the signs everywhere: enormous boulders stranded in fields, U-shaped valleys, rounded hills, gravel ridges, drumlins and moraines. At first no one believed him. Today scientists all agree that thousands of years ago great glaciers moved down from the Arctic four times and covered large parts of the Northern Hemisphere. Much the same thing happened in the Southern Hemisphere. The remains of these glaciers can still be seen in the Arctic and Antarctic regions.

GLACIERS LEAVE THEIR MARKS ON A LANDSCAPE

rounded hills

moraine

lake

ridges

drumlins

the earth through the ages

How old is the earth? It is hard to say exactly. However, geologists can find the age of some rocks by a complicated test for uranium. The oldest rock known goes back about three billion years. Geologists think, however, that the earth is older than that. It is probably at least four and one half billion years old.

Most geologists think the oceans and continents have always been in the same place as they are now, ever since the crust cooled enough for them to form. At times the land has lowered, however, and shallow seas have crept over parts of the continents. In those places thick layers of sedimentary rock have formed and covered the basement rocks. In the sedimentary rocks are many *fossils*.

Fossils are traces of ancient plants or animals which are preserved in the earth's crust—most often in sedimentary rock. This is how it happened. A plant or animal, when it died, may have been covered at once by water and a layer of mud or sand. This protected the plant or animal so that its hard parts—leaves or shells or teeth or bones—did not decay. As the sediments thickened and were pressed into rock, the plant or animal parts were pressed with them. Now, millions of years later, in places where the sedimentary rock has worn away, we find traces of them as fossils.

Sometimes they are the original shells or bones. Sometimes minerals have taken the place of the original material, and the fossils are exact mineral copies of the shells or bones or leaves. Or sometimes the material has disappeared altogether and left a hollow

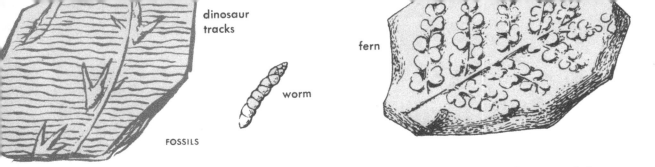

FOSSILS

mold. Animal footprints are sometimes preserved as molds.

We are lucky to have fossils. For without them we would know nothing about ancient life on the earth. Geologists use fossils and sedimentary rocks as clues in unraveling the earth's story.

One of the first to understand the value of these clues was a man who wasn't a geologist at all. He was a surveyor named William Smith. He lived in England in the 1700's, before anyone knew a great deal about fossils. As he traveled around on his surveying jobs, and especially as he worked with men who were digging canals, Smith became interested in rocks and fossils. Often the rocks were in layers or *strata*. Smith began to notice that certain kinds of fossil shells were always found together in one particular rock layer—but they were never found in the layer above or below it. However, the layer above had its special fossils, too, and so did the layer below.

As he went from place to place, Smith kept looking for these particular groups of fossils, and he noticed the layers they were in. After a while he saw that the fossils did not appear hit or miss, just anywhere in the rocks. Instead, particular kinds of fossils were found together and only in particular layers of rock. Smith decided that each of the creatures whose fossils were in a rock layer must have lived at one particular time—the period when that layer was forming.

He also noticed that the rock layers and fossils lay in a certain

sponge

coral

order. Suppose that there were three groups of fossils. X group, Y group, and Z group. X group of fossils always appeared in the lowest layer of the three. Y group of fossils was in the middle layer. Z group was in the top layer. X fossils must have lived first, thought Smith. They died and were covered by a layer with live fossils in it. Y fossils died and were followed by Z fossils.

Moreover, Smith found that in one place Y fossils might appear in red sandstone, and in another place they might appear in brown sandstone. In one place they might be buried under 16 layers of rock, and in another under only three. But the two sandstones, the red and the brown, were between the layers with X fossils and Z fossils. So, the red and brown sandstones must have been formed at the same time, when Y fossils were living, Smith reasoned.

After a while he found he could recognize rock layers by their fossils, no matter where the layers were or how deep they were buried. And he could also tell the age of one layer as compared with the age of another.

Finally, Smith began to make diagrams of the rock layers in various parts of England, sliced right down through as though they were layer cakes. On each layer he marked the kind of rock and the fossils he found in it. Now he could see where each layer was, in relation to all the other layers.

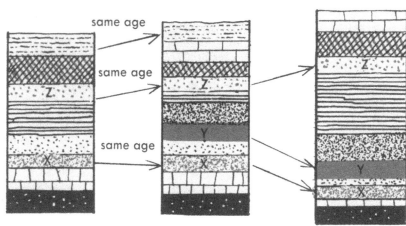

X, Y, and Z strata always occur in the same order in the rocks, even though sometimes one stratum may have been worn away and sometimes other strata may come between them.

People laughed at Smith for being so excited about rock layers, or strata. They called him "strata" Smith. But geologists realized how important Smith's diagrams were. Soon many geologists were making similar diagrams of other regions. Each diagram was a history of the earth at that particular place. Some of the rock layers were folded and broken. Some layers had almost worn away. Some seemed to be entirely missing. Between layers in some places there was lava from volcanoes. In other places there was igneous rock. Some layers revealed many fossils, some almost none.

By matching the diagrams of various regions, geologists could often tell what layer had worn away in one, because it still might be in its right place in another region nearby. They worked out the order in which the rock layers with their fossils were formed. Gradually they were able to piece together a huge time chart of the earth's past.

They found that they could divide their time chart into five large sections. At the beginning of each one there had been a time of great mountain building. Then gradually the earth had worn down, the seas had crept over parts of the land, and layers of sediment had formed. Millions of years later the land had again uplifted into mountains. And the wearing away had started all over again. The five large time sections are called *eras*. All their names end in *–zoic* (from the Greek word "life"). The Archeozoic ("beginning life") and the Proterozoic ("earlier life") are the oldest. They make up almost three-fourths of the whole age of the earth. There are almost no fossils in their rock layers, so we know very little about life in those times. The other

Mesozoic dinosaur, Tyrannosaurus

three eras are the Paleozoic ("ancient life"), the Mesozoic ("middle life"), and the Cenozoic ("recent life"). We live in the Cenozoic era.

Between the big mountain-building times there were smaller land changes. Geologists have used these to divide their time chart into *periods* and *epochs*. They know what rocks were formed in each, and what plants and animals lived then.

Fossils are our only clue to ancient life. Scientists called paleontologists study fossils and the rock layers where they are found. Sometimes the remains of whole animals appear in the rocks. More often there are a few bones here and a few bones there. Over many years paleontologists have worked, fitting the fossil clues together like part of a giant jigsaw puzzle. From what they have discovered we have a remarkably good idea of how life on earth developed. On the next page is the time chart the scientists have worked out.

MOUNTAIN-BUILDING

Period	North America	South America	Europe	Asia
RECENT				
PLEISTOCENE				Himalaya Mountains
PLIOCENE		Cascade Range		
MIOCENE				
OLIGOCENE	Rocky Mountains	Coastal Andes	Swiss Alps	
EOCENE				
PALEOCENE				
CRETACEOUS		Andes	Pyrenees	
JURASSIC	Sierra Nevada		Eastern Alps	
TRIASSIC		Palisades		
PERMIAN				
CARBONIFEROUS Mississippian and Pennsylvanian	Appalachian Mountains		Central Europe	
DEVONIAN	Acadian Mountains			
SILURIAN			Scottish Highlands	
ORDOVICIAN				
CAMBRIAN	Taconic Mountains			
PRE-CAMBRIAN	Laurentian Highlands / Adirondack Mountains			

As you can see, life started with very small forms in the water, followed by jellyfish. More tiny animals with shells came next, then coral and other creatures without backbones. Next came the fish, the earliest group of animals with backbones The first large group of animals to live on land at all were amphibians, but they had to spend part of their time in the water. Then came the reptiles, the first large group of full-time land animals with backbones. They included the giant dinosaurs. Gradually the reptiles gave way to the mammals, the animals who nursed their young with milk. Mammals became an important group of animals and remain so today.

Of all the mammals, man is the most numerous. He has changed the earth, possibly more than any river or any glacier ever did. He has plowed the soil for planting food, and cut down trees for fuel and building. He has dug the earth for minerals, built dams across the rivers for power, and watered dry, useless land for growing crops.

But man has not always been wise in his use of the earth. In his eagerness to take its treasures, he has worn it away too quickly. More and more we understand the damage that man has done. We realize that we must work to save the forests, the soil, the minerals, and the water. Each of us can help.

It took millions of years to make the earth as it is today. Think of that when you climb the little hill near your home. Perhaps that hill was once a mountain, and perhaps at some other time it lay beneath the sea. It had a beginning—long, long ago.

Index

Agassiz, Louis, 53,54
Asteroids, 2,3
Atmosphere, 3, 9, 10, 17, 39, 49
Atoms, 12-15
Batholiths, 34-35
Bridges, natural, 37, 46, 50
Canyons, 42-46
Caves, 49-51
Compound, 14, 15
Continents, 11, 55
Crust, earth's, 8-11, 15, 17, 19, 23-27, 29, 30, 55
    folding of, 24-26, 58
Delta, 41
Desert, 36
Dike, 34, 35, 44
Drumlin, 53, 54
Dune, 36, 38
Earth, age of, 55-62
Earthquakes, 27-28, 29
Elements, 14, 15, 16
Faults, 27
Folding, earth's crust, 24-26, 58
Fossils, 55-58
Fumaroles, 31
Geyser, 32-33
Glacier, 52-54
Grand Canyon, 43-36
Gravity, 3-4, 7, 21
Ground water, 47-51
Heat, 7, 8, 9, 10, 15, 16, 17, 20-21, 29-34
Hot springs, 32
Ice, 18, 19, 52-54
Igneous rock, 16-17
Laccoliths, 34
Lakes, 48, 53
Lava, 21, 29-30, 44-45, 48, 58
Layers, rock, 19-20, 24-26, 42, 44-45, 55, 56-58, 59
Lost rivers, 50
Magma, 20-21, 29-35
Metamorphic rocks, 20-21, 23

Meteorites, 9
Meteors, 2-3
Minerals, 15, 17, 21, 22,34, 49, 50, 51, 55
Moons, 2, 3, 4
Moraines, 53, 54
Mountains, 19, 23-26, 29-30, 34-35, 44, 52, 58-59
Ocean, 10, 11, 17, 23, 24, 25, 44-45, 55
Petrified material, 51, 55
Planets, 2-5
    beginning of, 6-7
Plateau, 24, 42, 45
Powell, John Wesley, 43-45
Pressure, rock, 8, 9, 20, 21, 22, 24-27, 29, 55
Rivers, 18, 19, 39-46, 50
Rock, 2, 7-12, 15-35
    igneous, 16-17
    metamorphic, 20-21, 23
    sedimentary, 20, 23-26, 42, 44-45, 55, 56-58
    wearing away of, 17-20, 35, 36-37, 40, 42, 44-46, 49-51, 53, 58
Sedimentary rock, 20, 23-26, 42, 44-45, 55, 56-58
Seismograph, 27-28
Sill, 34, 35
Sinkhole, 50, 51
Smith, William, 56-58
Springs, 32, 48
Stars, 3, 6
Steam, 29, 31-32, 33
Stata, 56-58
Sun, 2-5, 6, 7
Swamps, 48
Time chart, geologic, 58-62
Valleys, 39-41, 42-46, 52, 53, 54
Volcanoes, 29-30, 31, 44, 45, 58
Water, 10, 17, 19-20, 31-33, 36, 39-54
Waterfalls, 40, 43
Water table, 48
Wind, 20, 36-38, 45

CPSIA information can be obtained
at www.ICGtesting.com
Printed in the USA
BVHW011540250219
541082BV00018B/1811/P